A History of Shoemaking

Shoemaking, at its simplest, is the process of making footwear. Whilst the art has now been largely superseded by mass-volume industrial production, for most of history, making shoes was an individual, artisanal affair. 'Shoemakers' or 'cordwainers' (cobblers being those who repair shoes) produce a range of footwear items, including shoes, boots, sandals, clogs and moccasins – from a vast array of materials.

When people started wearing shoes, there were only three main types: open sandals, covered sandals and clog-like footwear. The most basic foot protection, used since ancient times in the Mediterranean area, was the sandal, which consisted of a protective sole, attached to the foot with leather thongs. Similar footwear worn in the Far East was made from plaited grass or palm fronds. In climates that required a full foot covering, a single piece of untanned hide was laced with a thong, providing full protection for the foot, thus forming a complete covering. These were the main two types of footwear, produced all over the globe. The production of wooden shoes was mainly limited to medieval Europe however – made from a single piece of wood, roughly shaped to fit the foot.

A variant of this early European shoe was the clog, which were wooden soles to which a leather upper was attached. The sole and heel were generally made from one piece of maple or ash two inches thick, and a little longer and broader than the desired size of shoe. The outer side of the sole and heel was fashioned with a long chisel-edged implement, called the clogger's knife or stock; while a second implement, called the groover, made a groove around the side of the sole. With the use of a 'hollower', the inner sole's contours were adapted to the shape of the foot. In even colder climates, such designs were adapted with furs wrapped around the feet, and then sandals wrapped over them. The Romans used such footwear to great effect whilst fighting in Northern Europe, and the native Indians developed similar variants with their ubiquitous moccasin.

By the 1600s, leather shoes came in two main types. 'Turn shoes' consisted of one thin flexible sole, which was sewed to the upper while outside in and turned over when completed. This type was used for making slippers and similar shoes. The second type united the upper with an insole, which was subsequently attached to an out-sole with a raised heel. This was the main variety, and was used for most footwear, including standard shoes and riding boots.

Shoemaking became more commercialized in the mid-eighteenth century, as it expanded as a cottage industry. Large warehouses began to stock footwear made by many small manufacturers from the area. Until the nineteenth century, shoemaking was largely a traditional handicraft, but by the century's end, the process had been almost completely mechanized, with production occurring in large factories. Despite the obvious economic gains of mass-production, the factory system produced shoes without the individual differentiation that the traditional shoemaker was able to provide.

The first steps towards mechanisation were taken during the Napoleonic Wars by the English engineer, Marc Brunel. He developed machinery for the mass-production of boots for the soldiers of the British Army. In 1812 he devised a scheme for making nailed-boot-making machinery that automatically fastened soles to uppers by means of metallic pins or nails. With the support of the Duke of York, the shoes were manufactured, and, due to their strength, cheapness, and durability, were introduced for the use of the army. In the same year, the use of screws and staples was patented by Richard Woodman. However, when the war ended in 1815, manual labour became much cheaper again, and the demand for military equipment subsided. As a consequence, Brunel's system was no longer profitable and it soon ceased business.

Similar exigencies at the time of the Crimean War stimulated a renewed interest in methods of mechanization and mass-production, which proved longer lasting. A shoemaker in Leicester, Tomas Crick, patented the design for a riveting machine in 1853. He also introduced the use of steam-powered rolling-machines for hardening leather and cutting-machines, in the mid-1850s. Another important factor in shoemaking's mechanization, was the introduction of the sewing machine in 1846 – a development which revolutionised so many aspects of clothes, footwear and domestic production.

By the late 1350s, the industry was beginning to shift towards the modern factory, mainly in the US and areas of England. A shoe stitching machine was invented by the American Lyman Blake in 1856 and perfected by 1864.

Entering in to partnership with Gordon McKay, his device became known as the McKay stitching machine and was quickly adopted by manufacturers throughout New England. As bottlenecks opened up in the production line due to these innovations, more and more of the manufacturing stages, such as pegging and finishing, became automated. By the 1890s, the process of mechanisation was largely complete.

Traditional shoemakers still exist today, especially in poorer parts of the world, and do continue to create custom shoes. In more economically developed countries however, it is a dying craft. Despite this, the shoemaking profession makes a number of appearances in popular culture, such as in stories about shoemaker's elves (written by the Brothers Grimm in 1806), and the old proverb that 'the shoemaker's children go barefoot.' Chefs and cooks sometimes use the term 'shoemaker' as an insult to others who have prepared sub-standard food, possibly by overcooking, implying that the chef in question has made his or her food as tough as shoe leather or hard leather shoe soles. Similarly, reflecting the trade's humble beginnings, to 'cobble' can mean not only to make or mend shoes, but 'to put together clumsily; or, to bungle.'

As is evident from this short introduction, 'shoemaking' has a long and varied history, starting from a simple means of providing basic respite from the elements, to a fully mechanised and modern, global trade. It is able to provide a fascinating insight not only into fashion, but society, culture and climate more generally. We hope the reader enjoys this book.

1882

Sept.	Sun.	Mon	Tue.	Wed.	Thu.	Fri.	Sat.
	...	...	...	...	...	1	2
	3	4	5	6	7	8	9
	10	11	12	13	14	15	16
	17	18	19	20	21	22	23
	24	25	26	27	28	29	30
Oct.	1	2	3	4	5	6	7
	8	9	10	11	12	13	14
	15	16	17	18	19	20	21
	22	23	24	25	26	27	28
	29	30	31	...	...	...	...

1882	Sun.	Mon.	Tue.	Wed.	Thu.	Fri.	Sat.
Nov.	...	...	...	1	2	3	4
	5	6	7	8	9	10	11
	12	13	14	15	16	17	18
	19	20	21	22	23	24	25
	26	27	28	29	30	...	...
Dec.	...	...	...	...	...	1	2
	3	4	5	6	7	8	9
	10	11	12	13	14	15	16
	17	18	19	20	21	22	23
	24	25	26	27	28	29	30
	31	...	...	...	...	...	...

1883

1883	Sunday	Monday	Tuesday	Wednes.	Thurs.	Friday	Saturd.
Jan.	...	1	2	3	4	5	6
	7	8	9	10	11	12	13
	14	15	16	17	18	19	20
	21	22	23	24	25	26	27
	28	29	30	31	...	...	...
Feb.	...	...	...	...	1	2	3
	4	5	6	7	8	9	10
	11	12	13	14	15	16	17
	18	19	20	21	22	23	24
	25	26	27	28	...	...	...
March	...	...	...	...	1	2	3
	4	5	6	7	8	9	10
	11	12	13	14	15	16	17
	18	19	20	21	22	23	24
	25	26	27	28	29	30	31
April	1	2	3	4	5	6	7
	8	9	10	11	12	13	14
	15	16	17	18	19	20	21
	22	23	24	25	26	27	28
	29	30	...	...	...	...	...
May	...	...	1	2	3	4	5
	6	7	8	9	10	11	12
	13	14	15	16	17	18	19
	20	21	22	23	24	25	26
	27	28	29	30	31	...	...
June	...	...	...	...	...	1	2
	3	4	5	6	7	8	9
	10	11	12	13	14	15	16
	17	18	19	20	21	22	23
	24	25	26	27	28	29	30

1883	Sunday	Monday	Tuesday	Wednes.	Thurs.	Friday	Saturd.
July	1	2	3	4	5	6	7
	8	9	10	11	12	13	14
	15	16	17	18	19	20	21
	22	23	24	25	26	27	28
	29	30	31	...	...	...	...
Aug.	...	...	...	1	2	3	4
	5	6	7	8	9	10	11
	12	13	14	15	16	17	18
	19	20	21	22	23	24	25
	26	27	28	29	30	31	...
Sept.	...	...	...	...	...	...	1
	2	3	4	5	6	7	8
	9	10	11	12	13	14	15
	16	17	18	19	20	21	22
	23	24	25	26	27	28	29
	30	...	...	...	...	...	...
Oct.	...	1	2	3	4	5	6
	7	8	9	10	11	12	13
	14	15	16	17	18	19	20
	21	22	23	24	25	26	27
	28	29	30	31	...	...	...
Nov.	...	...	...	...	1	2	3
	4	5	6	7	8	9	10
	11	12	13	14	15	16	17
	18	19	20	21	22	23	24
	25	26	27	28	29	30	...
Dec.	...	...	...	...	...	...	1
	2	3	4	5	6	7	8
	9	10	11	12	13	14	15
	16	17	18	19	20	21	22
	23	24	25	26	27	28	29
	30	31	...	...	...	...	...

1884

1884	Sun	Mo	Tue	Wed	Thu	Fri	Sat
Jan	...	...	1	2	3	4	5
	6	7	8	9	10	11	12
	13	14	15	16	17	18	19
	20	21	22	23	24	25	26
	27	28	29	30	31	...	...
Feb.	...	...	...	...	...	1	2
	3	4	5	6	7	8	9
	10	11	12	13	14	15	16
	17	18	19	20	21	22	23
	24	25	26	27	28	29	...

1884	Sun	Mo	Tue	Wed	Thu	Fri	Sat
Mar.	...	...	...	...	...	...	1
	2	3	4	5	6	7	8
	9	10	11	12	13	14	15
	16	17	18	19	20	21	22
	23	24	25	26	27	28	29
	30	31	...	...	...	...	...
Apr.	...	...	1	2	3	4	5
	6	7	8	9	10	11	12
	13	14	15	16	17	18	19
	20	21	22	23	24	25	26
	27	28	29	30	...	...	...

1884	Sun	Mo	Tue	Wed	Thu	Fri	Sat
May	...	...	...	...	1	2	3
	4	5	6	7	8	9	10
	11	12	13	14	15	16	17
	18	19	20	21	22	23	24
	25	26	27	28	29	30	31
June	1	2	3	4	5	6	7
	8	9	10	11	12	13	14
	15	16	17	18	19	20	21
	22	23	24	25	26	27	28
	29	30	...	...	...	...	...

→⁂Special⁜Notice.⁂←

WE present this Catalogue for the purpose of calling the attention of the public to some important facts regarding the manufacture of Boots and Shoes, hoping the information may prove of benefit to many, and increase the demand for the goods we manufacture. The main reason for there being so many inferior Boots and Shoes in the market is because consumers do not endeavor, or are unable, to inform themselves sufficiently regarding the quality of the goods they purchase, but are governed solely by appearance and price.

These are often unreliable tests, as inferior goods may be finished in an attractive style, and sold at high prices; the wearer soon experiencing to his regret the truth of the adage that "appearances are often deceitful."

Manufacturers of shoddy goods have been known to say that "the people will be humbugged, and we may as well reap the harvest as others."

For ourselves, we have to say that years of steady prosperity, and constant increase of business, has confirmed us in the theory (by which we have always been guided), that the majority of people are glad to patronize those manufacturers who endeavor to make honest goods that shall be worth all they cost. We ask this class to give our goods a trial.

C. M. HENDERSON & CO.

⇒ Past ✛ and ✛ Future. ⇐

OUR business which was established in a small way, in 1851, has been steadily increasing until we are sending out more goods annually than any other Boot and Shoe house in the world.

We have fully demonstrated the fact (in the face of the strongest competition) that good goods in in the end will bring success.

During the thirty years in which we have stood firmly to this principle, we have weathered every financial storm and panic, besides having been twice burned out with heavy loss; the increasing confidence of the public in the quality of our goods making it possible for us to withstand these misfortunes, and to go steadily on, enlarging our business to its present unequaled proportions.

With this experience we certainly shall not hesitate to continue on the same line—always making the best goods of their class in the market.

CHICAGO, ILLINOIS.

I N addition to the many important facts regarding the manufacture of

➤ BOOTS ✦ AND ✦ SHOES ◄

AND THE CHARACTER OF

OUR CUSTOM MADE GOODS,

We have given in this Catalogue

many valuable tables, rules, statistics,

etc., which will be found of interest,

besides being convenient for refer-

ence.

CHICAGO, ILLINOIS.

C. M. HENDERSON & CO., BOOTS AND SHOES, CHICAGO, ILLINOIS.

FIRST FLOOR, MAIN BUILDING.

OFFICES, SAMPLE ROOMS, AND MEN'S, BOYS' AND YOUTHS' FINE SHOE DEPARTMENT.

No. 1.
Hand-Sewed Button.
French or London Toe.
B, C, D and E Last.

No. 2.
Showing French Toe.

No. 3.
Hand-Sewed Dom Pedros.
French or London Toe.
C, D, and E Last.

No. 4.
Hand-Sewed and Pegged
Loop Lace Bal's.
French or London Toe.
C, D, and E Last.

No. 6.
Hand-Sewed Congress.
French or London Toe.
C, D, and E Last.

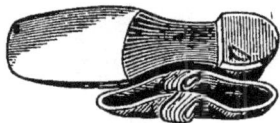

No. 7.
Showing London Toe.

No. 8.
Pegged Dom Pedros.
Wide and Medium Last.

No. 9.
Hand-Sewed Excelsior Tie.
London or French Toe.
C, D, and E Last.

No. 10.
Hand-Sewed French Tie.
French or London Toe.
C, D and E Last.

We make a large number of other hand-sewed
styles not represented here.

SECOND FLOOR, MAIN BUILDING.

WOMEN'S, MISSES' AND CHILDREN'S FINE SEWED SHOE DEP'T.

⇀TO THE LADIES.↼

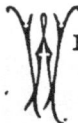

WE take occasion here to call your attention to our "BEST CUSTOM SEWED SHOES" for Ladies and Misses, some of the leading styles of which we represent by cuts on the opposite page.

We claim that for *Service, Comfort* and *Style,* they are not surpassed.

In the manufacture of all our goods, our first objective point is to insure *service,* and in our fine sewed shoes, as in our heavy boots, every shoe must have solid leather counters insoles, outsoles, taps and fillings, and all from the very best stock.

Comfort is the next point we aim at, therefore we have our shoes made on lasts as near the natural shape of the foot as possible, and with different widths, so that we can fit almost any foot that is not actually deformed.

Recognizing that ladies naturally admire the beautiful, we employ only the most competent and artistic draughtsmen, and therefore are able to excel other manufacturers in the style and symmetry of our *fine shoes,* and point to our "Opera" or "New York" Button or Side Lace in confirmation of this claim of superiority in *style.*

It will be found that a lady's foot will look smaller in one of our fine shoes than in almost any other.

Ask for "HENDERSON'S" BEST FINE SHOES, and see that our name is plainly stamped on the bottom, also on the inside lining of each shoe.

CHICAGO, ILLINOIS.

No. 11.
Common Sense Button.

No. 18.
Common Sense Polish.

No. 15.
Common Sense Last.

No. 12.
Opera Button.

No. 16.
New York Last.

No. 19.
New York Button.

No. 13.
Opera Polish.

No. 20.
New York Polish.

No. 17.
Opera Last.

No. 14.
New York Side Lace.

No. 21.
Opera Side Lace.

. No. 22.

Calf Sewed Button.
Women's, Misses' and
Children's.

No. 23.

Calf Sewed Polish.
Women's, Misses' and
Children's.

→Our Fine Calf Sewed Shoes←

EVERYONE should see the shoes represented above before buying this class of goods of any other make.

They are made after the most perfect fitting and stylish patterns, from the very best fine calf-skins, all solid, handsomely finished, and we know of nothing that we can add to make them better for the pnrpose intended,

Those who do not want to wear a pegged shoe for out-door heavy service, will find C. M. HENDERSON & CO'S FINE CALF SEWED SHOES the best thing in the market. See that our firm name is plainly stamped on each pair of shoes.

Ask your dealers to show them to you.

CHICAGO, ILLINOIS.

→How Boots and Shoes Should be Made←

NOW that *novelty* is often used as a magnet with which to draw money from the pockets of the public, and inventions multiply, we are frequently asked as to the best method of fastening the soles of Boots and Shoes to the uppers, and we unhesitatingly reply that PEGGING and SEWING excel all other methods.

Pegging cannot be equaled for economy, and should be used for the heavier goods, while sewing stands at the head for ease and pliability, and is best suited to the lighter and finer grades.

Wooden pegs are far more ECONOMICAL than metallic screws or nails; 1st, Because they cost vastly less, and will wear as long as the leather they are driven into; (whenever metallic fastened goods are pretended to be offered as cheaply as pegged, it may be assumed that a deception is being practiced, and that the screwed or nailed goods are really made of inferior stock.) 2d, Pegged goods can be repaired with ease and cheapness, while screwed or nailed boots and shoes are frequently thrown away when the soles are worn through, although the uppers may still be good, because of the difficulty and expense of getting them repaired.

There are several other serious objections to metallic fastened goods; one being that the metal is a conductor of heat and cold, and makes cold feet in winter; another is that the screws or nails are placed so far apart that the boots sometimes leak between the soles and uppers.

. A further objection, especially to screw fastenings, is that the screws frequently work upward through the insole and hurt the feet, *not being clinched;* also, when the outer sole is worn they often project, and are liable to catch in carpets, causing inconvenience and damage; so if persons desire to buy a metallic fastened boot or shoe, it is better to purchase those made with clinched nails. But the reasons above indicated are sufficient to show that either pegged or sewed goods are *much to be preferred.*

We, however, sell all kinds that the public demand, deeming our duty in the matter fully performed when we have made known the result of our experience and observation; but *advise* you to buy "HENDERSON'S" custom-made *pegged* or *sewed* goods.

THIRD FLOOR, MAIN BUILDING.

PACKING ROOM AND HEAVY PEGGED SHOE DEPARTMENT.

GREASE YOUR BOOTS.

HEAVY and medium Boots and Shoes should be thoroughly oiled once or twice a week while being worn. Those who will spend a little time and care in this direction will find that their boots or shoes will wear much longer than if left to dry out, as they otherwise will. In sandy soil this is of as much importance as in muddy sections. The space between the edge of the sole and the upper where they join often becomes filled with gravel, and the constant friction cuts the upper off ; thorough greasing in that space will prevent this. Tallow is excellent for leather, but should not be applied when very hot.

Boots and Shoes should never be placed near a hot fire, especially when they are wet; if not actually burned the life of the stock may be destroyed, rendering it liable to crack.

CHICAGO, ILLINOIS.

No. 21.
Two-Buckle Plow Shoes.

No. 27.
Men's Lace Plow Shoes.

No. 25.
Three-Buckle Plow Shoes.

No. 28.
Challenge Brogans.

No. 26.
Alaska Brogans

No. 29.
Competition Brogans.

No. 30.
Lumbermen's 3-buckle
Driving Shoes.

No. 34.
Lace Gussett Mining
Shoes.

No. 31.
Lumbermen's 2-Buckle
Driving Shoe.

No. 35.
2-Buckle Gusset Mining
Shoes.

No. 32.
Lace Double Gusset
Mining Shoe.

No. 36.
Hob-Nailed Brogans.

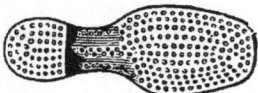

No. 33.
Showing the bottom of
our Nailed Mining
Shoes.

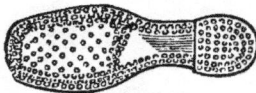

No. 37.
Showing the bottom of
our Hob-Nailed
Brogans.

No. 38.
Best Calf, Circle Seam
Pegged Polish.
Women's, Misses' and
Children's.

No. 41
Best Calf, Straight Seam
Pegged Polish.
Women's, Misses' and
Children's.

No. 39.
Calf Eureka Pegged Polish.
Women's, Misses' and
Children's.

No. 42.
Calf C. S. or S. S. Unlined
Pegged Polish.
Women's, Misses' and
Children's.

No. 40.
Kip Polkas.
Women's and Misses'.

No. 43.
Calf Pegged Polish.
New Style.
Women's and Misses'.

FOURTH FLOOR, MAIN BUILDING.

⇢ Boot ⟡ Department. ⇠

OUR "Best" Kip and Veal Calf Boots for Farmers' wear will be found the best in the market.

For twenty years we have been making these Boots here in Chicago, and we are willing to stand on the record they have made during that time.

We have hundreds of customers who say they will average a year's wear, and many say much longer. As a matter of economy you cannot do better than to buy

"HENDERSON'S" BEST BOOTS.

Our Fine Calf Boots, both French and American stock, Grain Boots, Plow Shoes, Brogans, Miners' Boots and Shoes, in fact everything made by us, will be found first-class in every re-spect. Look for the green label.

C. M. HENDERSON & CO.

CHICAGO, ILLINOIS.

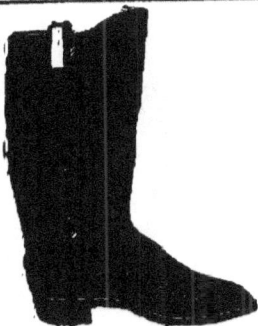

No. 44.
Best Kip Boots.
Russet Top.
Men's, Boys' and Youths'.

No. 47.
Best Kip Boots.
Plain Top.
Men's, Boys' and Youths'.

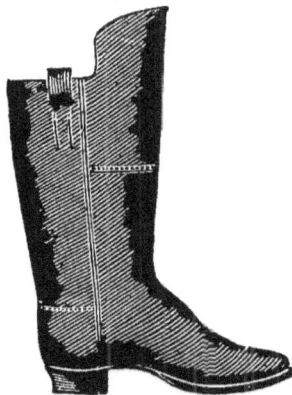

No. 45.
Western Star Kip Boots.
Men's, Boys' and Youths'.

No. 48.
Dress Veal Kip Boots.
Men's and Boys'.

No. 46.
Best Veal Calf Boots.
Men's,

No. 49.
Veal Kip Boots, Plain
Pieced Top. Men's.

No. 50
Best Grain Boots.
18-20 in. leg, ½D or Tap Sole.

No. 53
Grain Napoleon Boots.
½D or Tap Sole.

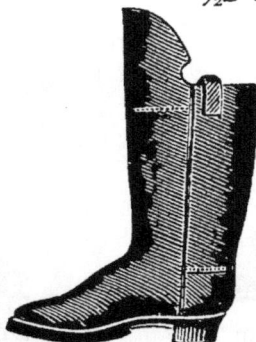

No. 52
Best Grain Boots.
16-18 in. Leg, ½D or Tap
Sole.

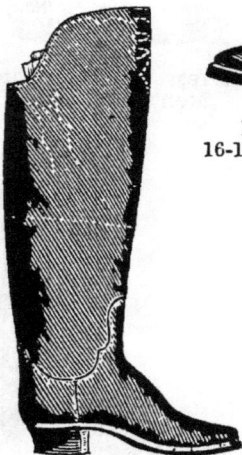

No. 51
Fine Grain Cavalry Boots.
18-22 in. Napoleon Leg.
Men's and Boys'.

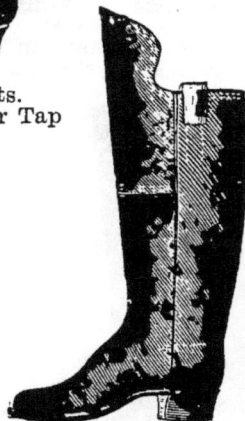

No. 54
Grain Pieced Top Boot.
16-18 in. Leg. } Men's & Boys.
14-16 in. Leg. }

CHICAGO, ILLINOIS.

No. 55.
Grain Plow Boots.
Single or Double Sole.

No. 56.
Grain Harvest Boots.
Double or Single Sole.

OUR Oil Grain Plow Boots are just the thing for farmers' summer wear. They are lighter and more pliable than a kip boot, and being wide on the bottom, with low, flat heels, are more comfortable in plowed ground than anyother style of boot we make.

No. 57
Lumbermen's Driving Boots.

No. 58.
Fine Calf Goat Leg
Opera Boots.
High, Medium and
Low Heel.

No. 61.
French and American Calf.
High, Medium and Low Heel.

No. 60
Surprise Calf.
High and Medium Heel.

No. 59.
Fine Calf Boots
Indestructible Tap Sole.

No. 62.
French Calf Sewed Boots
C, D, and E Last.

The following table shows the population of the
United States, according to the census of 1880.

STATE.	TOTAL.	WHITE.	COLORED.
Alabama	1,262,794	662,328	600,249
Arizona	40,441	35,178	138
Arkansas	802,564	591,611	210,622
California	864,686	767,266	6,168
Colorado	194,649	191,452	2,458
Connecticut	622,683	610,884	11,428
Dakota	135,180	133,177	381
Delaware	146,654	120,198	26,450
District of Columbia	177,638	118,236	59,378
Florida	267,351	141,832	125,464
Georgia	1,539,048	814,251	724,685
Idaho	32,611	29,011	58
Illinois	3,078,769	3,032,174	46,248
Indiana	1,978,362	1,939,094	38,998
Iowa	1,624,620	1,614,666	9,443
Kansas	995,966	952,056	43,096
Kentucky	1,648,708	1,377,187	271,461
Louisiana	940,103	455,007	483,794
Maine	648,945	646,903	1,418
Maryland	934,632	724,718	209,897
Massachusetts	1,783,012	1,764,004	18,411
Michigan	1,636,331	1,614,078	14,986
Minnesota	780,806	776,940	1,558
Mississippi	1,131,592	479,371	650,337
Missouri	2,168,804	2,023,568	145,046
Montana	39,157	35,446	288
Nebraska	452,433	449,806	2,376
Nevada	62,265	53,574	465
New Hampshire	346,984	346,264	646
New Jersey	1,130,983	1,091,947	38,796
New Mexico	118,430	108,127	648
New York	5,083,810	5,017,116	64,969
North Carolina	1,400,047	867,478	531,351
Ohio	3,198,239	3,118,344	79,665
Oregon	174,767	163,087	486
Pennsylvania	4,282,786	4,197,106	85,342
Rhode Island	276,528	269,931	6,503
South Carolina	995,622	391,224	604,275
Tennessee	1,542,463	1,139,120	402,991
Texas	1,592,574	1,197,499	394,001
Utah	143,906	142,380	204
Vermont	332,286	331,243	1,032
Virginia	1,512,806	880,981	631,754
Washington	75,120	67,349	357
West Virginia	618,443	592,606	25,806
Wisconsin	1,315,480	1,309,622	2,724
Wyoming	20,788	19,436	299
United States	50,152,866	43,404,876	6,577,151

White, 43,404,876; Colored, 6,577,151; Asiatic, 105,717;
Indians, 65,122.

CHICAGO, ILLINOIS.

FIFTH FLOOR, MAIN BUILDING.

SURPLUS STOCK AND SAMPLE PACKING ROOM.

BUY THE STARRETT STRAPS.

Sandals to wear in the mud, they will stay on the feet.

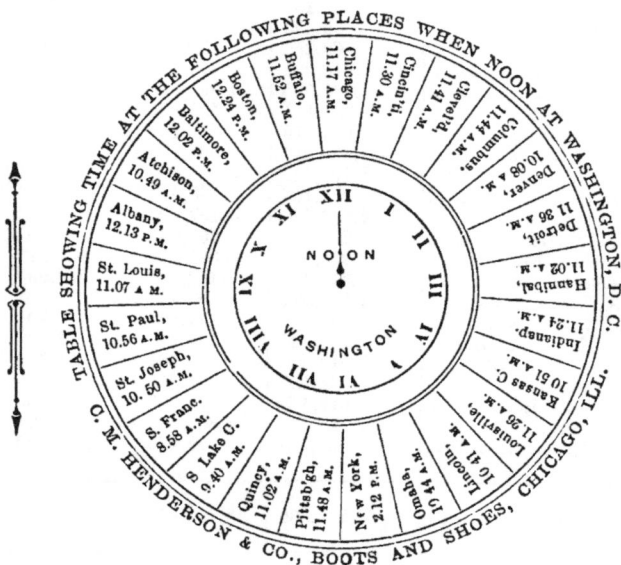

TABLE SHOWING TIME AT THE FOLLOWING PLACES WHEN NOON AT WASHINGTON, D. C.

C. M. HENDERSON & CO., BOOTS AND SHOES, CHICAGO, ILL.

NOON — WASHINGTON

Boston, 12.24 P.M.
Buffalo, 11.52 A.M.
Chicago, 11.17 A.M.
Cincin'ti, 11.30 A.M.
Cleve'd, 11.41 A.M.
Columbus, 10.08 A.M.
Denver, 11.36 A.M.
Detroit, 11.02 A.M.
Hannibal, 11.24 A.M.
Indianap., 10.51 A.M.
Kansas C., 11.36 A.M.
Louisville, 11.41 A.M.
Lincoln, 10.44 A.M.
Omaha, 2.12 P.M.
New York, 11.48 A.M.
Pittsb'gh, 11.02 A.M.
Quincy, 9.40 A.M.
S. Lake C., 8.58 A.M.
S. Franc., 10.50 A.M.
St. Joseph, 10.56 A.M.
St. Paul, 11.07 A.M.
St. Louis, 12.13 P.M.
Albany, 10.49 A.M.
Atchison, 12.02 P.M.
Baltimore,

CHICAGO, ILLINOIS.

Rubber Boots and Shoes.

AS we are dealing very largely in RUBBER BOOTS and SHOES, we wish to call the attention of the public to the importance of taking proper care of these goods, if they would secure the service reasonably to be expected from them.

Rubber Boots and Shoes were made originally to protect the feet in wet weather, or mud, and not to be worn in the house or during dry weather; they were afterwards made with warm linings, adapted to keep the feet warm, as well as dry, in extreme cold weather. But they are not economical for every day use, because the material is comparatively tender and not capable of enduring hard usage to the degree that leather is.

Rubber should never be brought near a hot fire, or kept in a very warm place when not in use. We venture to say that more Rubber Boots are destroyed by being subjected to heat than are worn out by actual service.

We are having made a new style *Rubber-Lined Boot*, which, for use in warm weather, will be found much better than the wool-lined, as they are cooler, do not cause the feet to perspire, and can readily be washed out at any time.

BASEMENT, MAIN BUILDING.

RUBBER BOOT AND SHOE DEPARTMENT.

On the opposite page we illustrate our

"Special Extra Quality Arctic".

AND OUR

"WARRANTED RUBBER BOOT."

This Arctic is better, not only in having an extra-heavy sole, but also in having an extra quality of upper, and it will average to outwear two pairs of ordinary Arctics.

Our "Warranted Boot" is unquestionably the best Rubber Boot made. It has an extra improved diamond tap, which is superior to any other tap made. If you want a Boot or Arctic that if properly used will wear you all winter, and with careful treatment two winters, ask for "HENDERSON'S EXTRA QUALITY ARCTIC," and "HENDERSON'S WARRANTED RUBBER BOOT."

CHICAGO, ILLINOIS.

No. 63
"Henderson's Special"
Extra Quality Arctics.

No. 64
Henderson's Warranted Rubber Boots.

FACTORY.—CUTTING ROOM.

Buy Henderson's Extra Arctics, the best ever made, and cheapest in the end.

POPULATION OF OUR LARGE CITIES.

The following Table shows the Population of the Cities of the United States having over 100,000 Inhabitants.

	1880.	1870.
New York, N. Y.	1,206,299	942,292
Philadelphia, Pa.	847,170	674,002
Brooklyn, N. Y.	566,663	396,099
Chicago, Ill.	503,185	298,997
Boston, Mass.	362,839	250,526
St. Louis, Mo.	350,518	310,864
Baltimore, Md.	332,313	267,354
Cincinnati, O.	255,139	216,239
San Francisco, Cal.	233,959	149,473
New Orleans, La.	214,090	191,418
Cleveland, O.	160,146	92,829
Pittsburg, Pa.	156,389	86,076
Buffalo, N. Y.	155,134	117,714
Washington, D. C.	147,293	109,199
Newark, N. J.	136,508	105,058
Louisville, Ky.	123,758	100,753
Jersey City, N. J.	120,722	82,546
Detroit, Mich.	116,340	78,577
Milwaukee, Wis.	115,587	71,440
Providence, R. I.	104,857	68,904

CHICAGO, ILLINOIS.

FACTORY.—FITTING ROOM.

PRESS COMMENTS.

(From the Chicago Tribune, Dec. 31, 1881.)

"It (the firm of C. M. Henderson & Co.) has been scrupulously careful that its goods should be just as represented, and it has been because their customers have always found their goods could be relied on, and that their promises never outran their performance, that their business grew constantly larger and larger, and the circle of their patrons became wider and wider, till their trade extended in all directions to the limit of Chicago enterprise, and for the past three years its sales have exceeded that of any boot and shoe house in the United States."

(Chicago Times, Dec. 31, 1881.)

"This house (C. M. Henderson & Co.) is probably selling now, and have been selling for several years, more boots, shoes and rubbers than any other jobbing house of the same line in the United States."

(The Inter Ocean, Dec. 31, 1881.)

"When we say that it (the firm of C. M. Henderson & Co.) does the largest jobbing trade in boots and shoes done by any house in any part of the world, we simply tell that which is known to well-informed merchants everywhere.

CHICAGO, ILLINOIS.

FACTORY.—SOLE LEATHER CUTTING ROOM.

⇒RULES FOR FINDING INTEREST⇐

On any Principal for any number of Days.   The answer being in Cents, separate the two right hand figures of
answer to express it in Dollars and Cents.

FOUR PER CENT.—Multiply the principal by the number of
days, cut off right hand figure from product, and divide by
nine.

FIVE PER CENT.—Multiply by number of days, and divide
by seventy-two.

SIX PER CENT.—Multiply by number of days, cut off right
hand figure and divide by six. (For 12 per cent., divide by
three instead of six.)

EIGHT PER CENT.—Multiply by number of days, and divide
by forty-five.

NINE PER CENT.—Multiply by number of days, cut off right
hand figure, and divide by four. (For 18 per cent., divide by
two instead of four.)

TEN PER CENT.—Multiply by number of days, and divide
by thirty-six. (For twenty per cent., divide by eighteen instead of thirty-six.)

FIFTEEN PER CENT.—Multiply by number of days, and
divide by twenty-four.

FACTORY.—BOTTOMING ROOM.

Buy the "Henderson" Warranted Rubber Boot.
Warranted not to crack.   The best ever made.

HOW TO LAY OFF A SQUARE ACRE OF GROUND.

Measure 209 feet on each side. 'and you will have a square acre, within an inch.

CONTENTS OF AN ACRE.

An acre contains 4,840 square yards.
A square mile contains 640 acres.

LAND MEASURE.

144 square inches,	1 square foot.
9 square feet,	1 square yard.
30¼ square yards,	1 square rod.
40 square rods,	1 square rood.
4 square roods,	1 square acre.
640 square acres,	1 square mile.

MEASURE OF DISTANCES.

A mile is 5,280 feet, or 1,760 yards in length.
A fathom is six feet.
A league is three miles.
A "Sabbath-day's journey" is 1,155 yards—(this is 18 yards less than two-thirds of a mile.)

A "day's journey" is 33½ miles.
A cubit is 2 feet.
A great cubit is 11 feet.
A hand (horse measure), is four inches.
A palm is three inches.
A span is 10½ inches.
A pace is three feet.

LENGTH MEASURE.

12 inches	1 foot.
3 feet	1 yard.
2 yards	1 fathom.
16½ feet	1 rod.
4 rods	1 chain.
10 chains	1 furlong.
8 furlongs	1 mile.
3 miles	1 league.

BARREL MEASURE.

A barrel of flour weighs 196 pounds.
A barrel of pork, 200 pounds.
A barrel of rice, 600 pounds.
A barrel of powder, 25 pounds.
A firkin of butter, 56 pounds.
A tub of butter, 84 pounds.

FACTORY.—HAND-SEWED GOODS DEPARTMENT.

If you want the best finished Calf Hand-Sewed Shoe made,
call for "Henderson's."

BUSINESS LAW CONDENSED.

The following brief compilation of Business Law is worth
a careful preservation, as it contains the essence of a large
amount of legal information.

It is not legally necessary to say on a note "for value
received."

A note by a minor is voidable at his option.

A contract made by a minor is voidable at his option.

A contract made with a lunatic is void.

A note obtained by fraud, or from a person in a state of
intoxication, in hands of original payer, cannot be collected.

If a note is lost or stolen, it does not release the maker;
he must pay it, if the consideration for which it was given
and the amount can be proven.

An indorser of a note is exempt from liability unless suit
be commenced against the maker on the first term after maturity, or unless such suit would be unavailing, or unless at
maturity of note, the maker is absent from the State so that
process could not be served.   This applies to Illinois only.

Notes bear interest from date only when so stated.

Principals are responsible for the acts of their agents
within the scope of their authority.

Each individual in a partnership is responsible for the
whole amount of the debts of the firm.

Ignorance of the law excuses no one.

It is a fraud to conceal a fraud.

The law compels no one to do impossibilities.

An agreement without consideration is void.

Signatures made with a lead pencil are good in law.

A receipt for money is not always conclusive.

The acts of one partner bind all the rest in all partnership
transactions.

CHICAGO, ILLINOIS.

FACTORY—FINISHING AND PACKING DEPARTMENT.

EVERY Boot or Shoe that goes out of our factory has the pegs properly cleaned out, saving both time and annoyance to the purchaser as well as the dealer.

→WEDDING CELEBRATIONS.←

Three Days.—Sugar.
Sixty Days—Vinegar.
1st Anniversary—Iron.
5th Anniversary—Wooden.
10th Anniversary—Tin.
15th Anniversary—Crystal.
20th Anniversary—China.
25th Anniversary—Silver.
30th Anniversary—Cotton.
35th Anniversary—Linen.
40th Anniversary—Woolen.
45th Anniversary—Silk.
50th Anniversary—Gold.
75th Anniversary—Diamond.

CHICAGO, ILLINOIS.

→✦Just✦for✦Fun.✦←

SIMILES.

As wet as a fish; as dry as a bone.
As live as a bird; as dead as a stone.
As plump as a partridge; as poor as a rat.
As strong as a horse; as weak as a cat.
As hard as flint; as soft as a mole.
As white as a lily; as black as coal.
As plain as a pikestaff; as rough as a bear.
As tight as a drum; as free as the air.
As heavy as lead; as light as a feather.
As steady as time; as uncertain as weather.
As hot as an oven; as cold as a frog.
As gay as a lark; as sick as a dog.
As slow as a tortoise; as swift as the wind.
As true as the gospel; as false as mankind.
As thin as a herring; as fat as a pig.
As proud as a peacock; as blithe as a grig.
As savage as tigers; as mild as a dove.
As stiff as a poker; as limp as a glove.
As blind as a bat; as deaf as a post.
As cool as a cucumber; as warm as toast.
As flat as a flounder; as round as a ball.
As blunt as a hammer; as sharp as an awl.
As red as a ferret; as safe as the stocks.
As bold as a thief; as sly as a fox.
As straight as an arrow; as crooked as a bow.
As yellow as saffron; as black as a sloe.
As brittle as glass; as tough as gristle.
As neat as my nail; as clean as a whistle.
As good as a feast; as bad as a witch.
As light as the day; as dark as pitch,
As wide as a river; as deep as a well.
As still as a mouse; as loud as a bell.
As sure as a gun; as true as a clock.
As frail as a promise; as firm as a rock.
As brisk as a bee; as dull as an ass.
As full as a tick; as solid as brass.
As lean as a greyhound; as rich a Jew.
And ten thousand similes equally new.

AGRICULTURAL STATISTICS.

Below will be found the number of bushels of three of the staple grains of the United States, produced in 1879, according to the census report. The estimates for 1881 are not complete at this writing, but show the corn crop to be about 1,150,000,000 bushels, wheat about 400,000,000 bushels.

STATES AND TERRI-TORIES.	CORN, BUSHELS.	WHEAT. BUSHELS.	OATS, BUSHELS.
Alabama	25,446,413	1,529,683	3,039,274
Arizona	36,246	189,527	624
Arkansas	23,666,057	1,252,181	2,187,777
California	2,050,007	28,787,132	1,355,871
Colorado	455,988	1,475,559	640,100
Connecticut	1,924,794	38,742	1,009,706
Dakota	2,078,089	3,018,354	2,331,230
Delaware	3,892,464	1,175,182	378,508
Dist. of Columbia	29,750	6,402	7,440
Florida	3,174,234	513	468,122
Georgia	23,190,472	3,158.335	5,544,161
Idaho	16,408	540,564	462,236
Illinois	327,796,895	51,130,455	63,206,250
Indiana	117,121,915	47,288,989	15,606,721
Indian Territory	....	....	....
Iowa	276,093,295	31,177,225	60,612,141
Kansas	106,791,482	17,324,141	8,180,385
Kentucky	73,977,829	11,355,340	4,582,968
Louisiana,	9,878,024	5,044	229,850
Maine	960,633	665,714	2,265,575
Maryland	16,202,521	8,004,484	1,794,872
Massachusetts	1,805,295	15,818	645,169
Michigan	36,844,229	35,537,097	18,190,493
Minnesota	14,979,744	34,625,657	23,372,752
Mississippi	21,340,800	218,890	1,959,620
Missouri	203,464,620	24,971,727	20,673,458
Montana	5,794	409,688	900,915
Nebraska	65,785,572	13,846,742	6,555,565
Nevada	12,891	70,404	186,860
New Hampshire	1,358,625	169,316	1,018,006
New Jersey	11,247,402	1,901,739	3,710,808
New Mexico	650,954	708,778	157,437
New York	26,520,182	11,586,754	37,575,506
North Carolina	27,959,894	3,385,670	3,830,622
Ohio	112,681,046	46,014.869	28,664,505
Oregon	127,675	7,486,492	4,393,593
Pennsylvania	47,970,987	19,462,405	33,847,439
Rhode Island	372,967	290	159,339
South Carolina	11,764,349	962,330	2,715,445
Tennessee	62,833,017	7,331,480	4,722,938
Texas	28,846,073	2,555,652	4,868,916
Utah	164,244	1,167,268	417,938
Vermont	2,022,015	337,257	3,742,282
Virginia	29,102,721	7,822,354	5,333,081
Washington T'y	39,906	1,921,382	1,581,951
West Virginia	14,233,799	4,002,017	1,908,505
Wisconsin	35,991,464	24,884,689	32,911,246
Wyoming	65	4,762	22,512
Totals, 1879	1,772,909,846	459,591,093	407,970,712
" 1880	1,717,434,543	498,549,868	417,885,380

IN our very extensive business, we are obliged to sell all classes of goods, but our aim has always been to encourage the sale of the best grades, because we believe this to be for the best interest of our customers, and we are certain it will be for your advantage to use for yourself and family none but

Henderson's Best Boots and Shoes

Merchants who may be handling our goods in your vicinity will be glad to show them to you.

Our name on any grade of boots and shoes is a recommendation, indicating that although they may not be our best grade, still they are honestly made and worth all they cost.

Our *very best boots* and *heavy shoes* have the *green label*, and our *best fine shoes* have our name stamped on the inside lining as well as on the bottom of each pair.

C. M. HENDERSON & CO.

CHICAGO, ILLINOIS.

PRESIDENTS OF THE UNITED STATES.

NAME.	WHEN BORN.	INAUGURAT'D	DIED.
Geo. Washington...	Feb. 22, 1732	April 30, 1789	Dec. 14, 1799
John Adams.........	Oct. 19, 1735	Mar. 4, 1797	July 4, 1826
Thomas Jefferson..	April 2, 1743	" 4, 1801	July 4, 1826
James Madison.....	Mar. 16, 1751	" 4, 1809	June 28, 1836
James Monroe.....	April 28, 1758	" 4, 1817	July 4, 1831
John Q. Adams.....	July 11, 1767	" 4, 1825	Feb. 23, 1848
Andrew Jackson...	Mar. 15, 1767	" 4, 1829	June 8, 1845
Martin Van Buren	Dec. 5, 1782	" 4, 1837	July 24, 1862
Wm. H. Harrison..	Feb. 9, 1773	" 4, 1841	April 4, 1841
John Tyler............	Mar. 2'', 1790	April 5, 1841	Jan. 17, 1862
James K. Polk......	Nov. 2, 1795	Mar. 4, 1845	June 15, 1849
Zachary Taylor ...	Nov. 24, 1784	" 4, 1849	July 9, 1850
Millard Fillmore...	Jan. 7, 1800	July 10, 1850	Mar. 8, 1874
Franklin Pierce....	Nov. 23, 1804	Mar. 4, 1853	Oct. 8, 1869
James Buchanan...	April 22, 1791	" 4, 1857	June 1, 1868
Abraham Lincoln..	Feb. 12, 1809	" 4, 1861	April 15, 1865
Andrew Johnson...	Dec. 29, 1808	April 15, 1865	July 31, 1875
Ulysses S. Grant....	April 27, 1822	Mar. 4, 1869	....
Ruth. B. Hayes......	Oct. 4, 1822	" 4, 1877	....
James A. Garfield..	Nov. 19, 1831	" 4, 1881	Sept. 19, 1881
Chester A. Arthur.	Oct. 5, 1830	Sept. 20, 1881	....

ELECTORAL VOTE of the SEVERAL STATES.

STATES.	1883	1880	STATES.	1883	1880
Alabama.................	10	10	Missouri...................	16	15
Arkansas................	7	6	Nebraska...................	5	3
California..............	8	6	Nevada....................	3	3
Colorado.................	3	3	New Hampshire	4	5
Connecticut..	6	6	New Jersey	9	9
Delaware................	3	3	New York...............	36	35
Florida....................	4	4	North Carolina........	11	10
Georgia..................	12	11	Ohio.....................	23	22
Illinois...................	22	21	Oregon....................	3	3
Indiana...................	15	15	Pennsylvania..........	30	29
Iowa.....................	13	11	Rhode Island...........	4	4
Kansas...................	9	5	South Carolina........	9	7
Kentucky...............	13	12	Tennessee.................	12	12
Louisiana.........	8	8	Texas........	13	8
Maine.	6	7	Vermont.	4	5
Maryland...............	8	8	Virginia	12	11
Massachusetts.........	14	13	West Virginia..........	6	5
Michigan	13	11	Wisconsin................	11	10
Minnesota.....	7	5			
Mississippi..............	9	8	Total...............	401	369

Deducting two from each State gives the number of Representatives in Congress.

9 781473 338203